# THE DIGITAL AGE: A JOURNEY THROUGH TIME

Chris Hughes

*I dedicate this book to the two greatest educators I ever knew, Jim Fish and Tony Read. Their unwavering dedication, inspiring lessons, and the personal guidance they provided have not only shaped my educational journey but have left an indelible mark on my life. This book is a reflection of the wisdom and values they instilled in me, and for that, I am eternally grateful.*

# INTRODUCTION

Welcome aboard our time-traveling tour of tech history, from the room-sized mainframes of the 1950s to the quantum wonders of the 2020s. This isn't just a tale of circuits and code; it's a saga of human cleverness and the ever-persistent 'have you tried turning it off and on again?' philosophy.

We'll zip through decades of digital dazzle, from the early computing giants (which doubled as unofficial room heaters) to the invasion of personal computers (far more welcome than garden gnomes). We'll also leap into the world of the Internet, mobile tech, and AI – where emojis replaced words, Zoom meetings became the new boardroom, and virtual reality often outshone, well, actual reality.

But this isn't just a historical jaunt; it's a peek into a future filled with AI butlers and quantum coffee machines. So, whether you're a tech enthusiast, a student of the past, or someone still baffled by the speed of broadband, there's a byte-sized treat for you here. Let's set off on this whimsical ride through the digital age, where technology has gone from being a luxury to our everyday sidekick.

# CONTENTS

# CHAPTER 1: THE 1950S - DAWN OF THE DIGITAL AGE

So, here we are, dear reader, stepping into the 1950s. Can you smell the new age dawning? Well, perhaps not with all those not-so-efficient petrol engines around. But, let's not get sidetracked – you're here for the thrilling world of vacuum-tube computers!

These technological whales, don't be fooled by their size, were pivotal in the annals of computer history. This decade laid the foundation of the digital age. Sure, they didn't have iPads or Steam Decks, but what they lacked in portability, they made up for in sheer guts – and size. Have you seen how enormous these things were? The UNIVAC and IBM Mainframes weren't just big; they were colossal!

But enough frame shaming! it's not about size – it's about capability. The IBM machines were in their element in the labs, crunching complex scientific calculations faster than ever. Both UNIVAC and IBM mainframes could process data at unprecedented speeds, making even the most savvy human calculators look, well, a bit less savvy.

Though the 1950s may not have had a headline event like the space race (keep an eye out for Chapter 2), it was nonetheless a groundbreaking era. These years set the stage for a computing revolution like never before. It was no longer about what these machines couldn't do; it was about the doors they would open, the possibilities they would unlock for humanity.

# CHAPTER 2: 1960S
# - THE TRANSISTOR
# REVOLUTION

Enter the 1960s, an era sandwiched between The Beatles, the space race, and "To Kill a Mockingbird." Amidst this cultural backdrop, something quite remarkable was born. This wasn't your ordinary birth; oh no,no,no! no flesh and bones here, but circuits and transistors instead. Rising above its antiquated predecessors, shedding the cumbersome vacuum tubes, the 'Atlas' computer emerged.

Now, this machine was colossal – think of a bird's nest of tubes and wires, resembling what a novice PC builder might end up with after enthusiastically attaching every single cable to a fully modular power supply.

Despite its size, it was what Atlas represented that truly mattered. Those beastly swats over there in the UK had kick-started humanity on a path brimming with amazing advances (and yes, the occasional strange blue screen, but let's not dwell on that). The future of computing was afoot, and Atlas was leading the charge!

From that moment, competition sprang up everywhere. From the UNIVAC to the IBM Mainframe, everyone wanted to be at the forefront of this new digital frontier.

As the first heartbeat of the digital age echoed through those halls, we made ourselves a stiff cup of tea patted ourselves on the back and congratulated ourselves, maybe a bit too much!

So, as humans are often inclined to do after finishing their tea, we took a good look around to see how we could outdo the Joneses down the street. Over at the White House, there was a pressing need to win the space race, and they had a bold idea: let's put humans in a tin can, attach a bunch of explosive compounds, and see what happens. To some, this seemed like a fantastic plan, but to make it even spicier, they decided to throw a computer into the mix. After all, who wants to be bogged down with calculations when you could be gazing out of a spacecraft window, am I right?

Enter Project Gemini – well, more specifically, the Gemini computer. This nifty device was a bit more compact than its predecessor, Atlas, yet still packed a punch, tipping the scales at nearly 27 kilograms. With a confident "She'll be right, mate," they lit the fuse, and off they went, with little Gemini bravely leading the way.

To the astonishment of many, Gemini passed every test with flying colours, leaving those vacuum-tube relics in its digital dust. This marked the beginning of a new era: the computer had stepped out of the lab and was now making its way into a broader, more exciting world!

# CHAPTER 3: 1970S
# - THE BIRTH
# OF PERSONAL
# COMPUTING

Well, the 1970s rolled in with quite a bang. The Beatles were on their way out, the Watergate Scandal was grabbing headlines, and can you believe a man actually died at his own funeral? Yeah, the 1970s were something else. But amid all this, the decade gifted us something monumental – the birth of personal computers!

After the swinging sixties, it was only a matter of time before those oversized calculators found their way onto the desks of eager teenagers. And what sparked this burgeoning industry? Was it climate change modelling? Delving into molecular biology? Nope – it was gaming! Classics like Pong, Space Invaders, and the unforgettable Adventure! Games became the rage, and if a machine could run them, you bet it was in high demand.

1975 was a game-changer with the introduction of the formidable Altair 8800. Alright, it wasn't in every household – it was pricey and a bit of a challenge to use. The concept of a GUI, or even a screen, was still in the realm of dreams. But those dreams were about to become a reality (no spoilers, I promise).

Come 1977, and Apple threw its hat in the ring with the Apple II. This machine was a revelation! It boasted a relatively user-friendly interface and, get this, could run multiple applications. Mind-blowing, right?

The 1970s weren't just about the hardware; it was an era that shifted mindsets. Sure, Uncle Billy might still be in denial about the need for computers in every home, but this decade planted a seed. A seed that would grow into a magnificent digital world, forever altering the landscape of humanity. The stage was set, and the world was poised for a digital revolution.

# CHAPTER 4: 1980S - THE AGE OF HOME COMPUTING

The 1980s were a whirlwind of action, both inside and outside the world of computing. While events like Mount St. Helens' eruption and John Lennon's assassination shook the world, the rise of home computing was making waves of its own. And let's not forget the release of Pac-Man – an absolute game-changer!

IBM wasn't about to let Apple hog the limelight. They entered the fray with the IBM Personal Computer (PC), and boy, did it revolutionize the personal computing scene. This machine opened up the market like never before, making home computers accessible to the masses. And the arrival of "PC clones" only added fuel to the fire, making it easier for everyone (especially persistent teenagers) to get their hands on a computer.

But Apple wasn't down for the count. In 1984, they bounced back with the Apple Macintosh. This machine wasn't just another computer; it introduced users to a graphical user interface (GUI), a monumental leap that made computing more intuitive and user-friendly.

Now, we've talked a lot about the hardware of the '80s, but that's not all it was known for. This decade also saw the birth of an operating system from a little company you might have heard of – Microsoft. Their initial OS brought GUIs to a wide range of IBM-compatible PCs, opening a world of possibilities. Office workers could write, parents could budget, and teenagers could – well, let's be honest – game to their heart's content.

And then there was this other little thing, barely more than a whisper at the time – the internet. Imagine, sending mail from your computer to someone in another city, all through this mysterious digital network. It sounded like a wild dream, but who knew it would become the cornerstone of our digital lives?

# CHAPTER 5: 1990S
# - THE INTERNET
# REVOLUTION

As we zoom closer to the modern age (or maybe I'm just trying to keep up!), the 1990s whirls into view with a smorgasbord of world-changing events. From the reunification of Germany to the fall of the Soviet Union, and let's not forget Pakistan clinching the cricket World Cup – it was a decade of monumental happenings. And the world of computing? Oh, it kept pace with the whirlwind.

The 1990s could easily claim the title of the most transformative decade in computing history, primarily due to the development and popularization of the Internet. Who could forget the iconic, high-pitched screech of a dial-up modem or the frustration of being kicked offline mid-StarCraft victory because someone picked up the phone? Those were the days of exploring the web and accidentally spoiling six months of Coronation Street in one sitting.

But we can't talk about the '90s without mentioning Windows 95. This software was a game-changer, firmly establishing Microsoft as a titan in the home operating system arena. It made computers friendlier, more versatile, and an essential part of the home.

The '90s also witnessed the dawn of the mobile phone revolution. Those massive circuits from the '70s? They were now small enough to fit in your hand, encased in a little plastic box we called a mobile phone. This was a game-changer in how we communicate and interact with the world.

As the decade drew to a close, reflecting on the journey from those humble beginnings was nothing short of astounding. By the end of the '90s, even my mum could shop online – a true testament to how integral computers had become in our daily lives. The Internet had transformed communication, making it easier than ever (and yes, it also gave us multiplayer gaming and a whole new way to shirk responsibilities – but let's call that a feature, not a glitch, right?).

# CHAPTER 6: 2000S
# - THE MOBILE AND
# SOCIAL ERA

Ah, the 2000s, stepping out of the millennium bug scare into a world that was, well, still standing. Sure, Y2K might have been a bit overhyped, but it didn't stop some savvy folks from cashing in. Amid significant events like September 11 and Hurricane Katrina, the world of computing was buzzing with its own groundbreaking developments.

This decade brought us the gift (or curse, depending on who you ask) of texting. Gone were the days of actual phone conversations; now, we were tapping away in ALL CAPS, limited to 160 characters. Who knew we'd become such masters of shorthand?

Then, in 2007, something called an "iPhone" burst onto the scene. Rumour has it, Apple's alphabet page got eaten by their dog, and they were left with just the letter 'i', hence the name (okay, I might be making that part up). This iPhone wasn't just a phone; it was a camera, an internet portal, and a selfie machine. Suddenly, everyone was snapping pictures of themselves and their meals, convincing themselves that the whole world was watching.

To capitalize on this selfie and status update craze, a little website named "Facebook" emerged and, boy, did it take off. It became the hub for endless streams of selfies, food photos, and an avalanche of friend requests.

And let's not forget the monumental shift from the screechy 56k modems to the glory of broadband. No longer did we have to choose between surfing the web and using the phone – we could do both!

The 2000s didn't just change mindsets; they revolutionized them. E-commerce boomed, allowing us to get everything done without moving more than our fingertips. What a time to be alive! Though, perhaps, the whole 'moving' part might have been more important than we realized.

# CHAPTER 7: 2010S
# - THE CLOUD
# AND AI ERA

Welcome to what some might call the 'modern era' – a time that was as eventful as it was innovative. Between Brexit shaking up Europe and Occupy Wall Street making headlines, the world of computing was making some groundbreaking strides of its own. Enter 'Cloud Computing' – a game-changer that revolutionized our digital lives. Suddenly, our entire digital existence could travel with us, making businesses more seamless and technology more flexible than ever.

But of all the leaps we've seen, including hurtling humans into space, the advent of AI was perhaps the most monumental. This wasn't just a technological shift; it was a fundamental change in how we accessed information and managed our daily lives.

Virtual assistants like Alexa, Google, and Siri (yes, even Siri) began to transform our homes into something akin to sci-fi novels. Smart devices proliferated, offering unprecedented convenience – except, of course, on those frosty mornings when my phone was charging in another room, and Google couldn't hear my desperate pleas for the weather forecast.

Our lives became intertwined with AI in ways we never imagined. It guessed our shopping needs, curated our music playlists, and even recommended movies to stream – and speaking of streaming, physical media? That was so last decade. Streaming was the new norm, and owning DVDs or CDs was quickly becoming a quaint relic of the past.

With the world more connected than ever, you'd think we'd all come together to sort out our differences. Well, not quite. The sabre-rattling among global powers continued, but in the realm of technology, we moved from leaps to bounds. The 2010s weren't just about new gadgets and software; they were about a cultural and societal transformation, driven by the power of technology.

# CHAPTER 8: 2020S
# - THE QUANTUM
# LEAP AND BEYOND

Here we are at the outset of the 2020s, and already this decade is shaping up to be a technological whirlwind, surpassing the combined advancements of previous eras.

Quantum computing has burst onto the scene, and while it might not yet be able to locate my perpetually misplaced car keys, it's revolutionizing our understanding of the universe. We're talking about tackling monumental challenges here – from unravelling medical mysteries to addressing climate change and deciphering the complex world of cryptography. These were once daunting tasks for traditional computing, but now, they're just another day at the office for quantum computers.

Then there's AR and VR – Augmented Reality and Virtual Reality. Once the darlings of the gaming world, they've rapidly evolved into powerful tools extending far beyond entertainment. We're seeing them transform education, revolutionize remote work, and offer new forms of immersive experiences that were mere fantasies a few years ago.

But let's address the mammoth in the room: Artificial Intelligence (AI). AI stands apart from its technological brethren in its sheer potential and impact. It's redefining our relationship with technology, intertwining with every aspect of our lives. From managing smart homes to transforming entire industries, AI's footprint is both vast and deep.

As we stand at the dawn of 2024, it's clear we're not just entering a new decade; we're stepping into a new epoch. Welcome to an era where the boundaries of technology are only limited by our imagination. And don't worry about mundane tasks like letting the cat out – our ever-helpful AI companions have got it covered!

# EPILOGUE

As we close this journey through the decades of computing, it's worth pausing to reflect on the profound impact these machines have had on our lives. From the hulking mainframes of the 1950s to the quantum computers and AI of the 2020s, each leap in technology has brought us closer to a world once imagined only in science fiction.

But what does the future hold? If history has taught us anything, it's that the realm of computing knows no bounds. We stand on the brink of possibilities that are as exciting as they are daunting. Quantum computing could solve problems we haven't even encountered yet, while AI might redefine our understanding of intelligence itself.

Yet, amidst this awe-inspiring progress, we must also be mindful of the challenges. Issues like data privacy, cybersecurity, and the digital divide remind us that with great power comes great responsibility. As we integrate technology even more deeply into our lives, we must navigate these waters with care, ensuring that our digital future is as inclusive and safe as it is innovative.

As we look to the horizon, one thing is clear: our journey with technology is far from over. It's a partnership that will continue to evolve, challenge, and inspire us in ways we can only begin to imagine.

So, dear reader, as you set this book aside, remember that you are part of this incredible story. The digital age is not just a tale of circuits, code, and silicon. It's a story of human curiosity, ingenuity, and the unending quest to push the boundaries of the possible.

# ENDNOTE REVIEW REQUEST

Thank you for reading my book! If you enjoyed it, I would be grateful if you could take a moment to leave a review on Amazon.

Your feedback is not only valuable to me, but it also helps new readers discover my work. I read and appreciate every review.

# ABOUT THE AUTHOR

## Chris Hughes

Hailing from a quaint town nestled in the deepest recesses of New Zealand, a place where dreams stretch as wide as the starlit skies, I was a young boy who dared to envision a world of discovery. Inspired by the echoes of an old beloved book series, 'Tell Me Why,' I embarked on a quest fueled by curiosity and wonder.

Now a man, I have journeyed far beyond the rolling hills of my youth, carrying with me a mission: to make the life sciences accessible and engaging for younger readers. My passion is not just for the mysteries of nature but for weaving them into the modern, bustling era.

With each page turned, I aim to ignite a spark of curiosity in young minds, encouraging them to look beyond the ordinary and to explore the extraordinary wonders of the world around them. My writings are more than just words; they are invitations to embark on adventures in learning, to uncover the marvels of biology, astronomy, and beyond.

As an author, I strive to ignite curiosity in the inquisitive minds of the next generation, guiding them through stories that are not only educational but also resonate with the spirit of exploration and discovery. My books await those who dare to ask, 'Why?'

# BOOKS BY THIS AUTHOR

## Abigail's Algorithm: The Quest Through Codeville

In Abigail's Algorithm: The Quest Through Codeville, Abigail goes on a digital adventure after activating Byte, a small robot, in her father's shed. Her journey through the world of Codeville introduces her to various programming languages, each unlocking new challenges and skills. From organizing data with COBOL to programming a light show using JavaScript, Abigail harnesses the power of coding to rescue her father. This captivating tale blends the excitement of discovery with the joys of learning to code, inspiring young readers to explore the endless possibilities of technology.

## Nora's Nuclear Journey: Adventures Of An Atom

Nora's Nuclear Journey: Adventures of an Atom is a story about Nora, a helium atom born in a distant star through nuclear fusion. Her journey begins with her being flung into space after her star explodes in a supernova, leading her to Earth. Here, Nora learns about her atomic structure and is introduced to the periodic table by scientists. She visits a nuclear power plant, gaining insights into nuclear energy and fission. Nora discovers the benefits and responsibilities of nuclear power, including safety and waste management. She befriends atoms involved in various energy sources, understanding the importance of a balanced energy mix. The story concludes with Nora reflecting on her cosmic journey, emphasizing the significance of energy, science, and our place in the universe.

## Leo's Light: A Journey Through Photosynthesis

Leo's Light: A Journey Through Photosynthesis takes readers on a magical journey with Leo, a photon born in the sun. Eager to explore, Leo travels across space to Earth, where he lands on a lush, green leaf. There, he meets Chloro, a chlorophyll molecule, and learns about the incredible process of photosynthesis. As Leo aids in converting sunlight into energy, he discovers the vital role photons play in creating food for plants and oxygen for all living creatures. This enchanting story beautifully illustrates the cycle of life, the importance of sunlight, and the wonders of nature. Perfect for young readers, it's a vivid exploration of science and the natural world.

## Astrid's Odyssey: A Microscopic Tale Of Survival And Defense

Astrid's Odyssey is not just a journey of a bacterium but a journey into the marvels of the human body's defense mechanisms. Through engaging storytelling and scientifically accurate descriptions, this book offers readers of all ages a unique perspective on how our bodies protect us from microscopic invaders. It's a tale that weaves together science, wonder, and the eternal dance of survival, making it a perfect read for anyone fascinated by the miracles of biology and the unseen battles that rage within us every day.

www.ingramcontent.com/pod-product-compliance
Lightning Source LLC
Chambersburg PA
CBHW050918290526
45792CB00002B/803